Steven [illegible]

Sigmar & Reini
with love
from Erika
2014

The mission of No Guff Press is to empower gardeners with practical, No Guff™ gardening advice.

Grow Figs Where You Think You Can't

www.Grow-Figs.com

Guff-Detecting, Word-Wrangling Editor: Mary Fran McQuade

Figgin Great Design: Bryan McCrae, Feedlot Studios

No Guff Press

www.NoGuffPress.com

Library and Archives Canada Cataloguing in Publication data available upon request to the publisher.

Printed in Canada

ISBN: 978-0-9868144-1-9

NO GUFF PRESS™

Grow Figs Where You Think You Can't

www.Grow-Figs.com

Steven Biggs

Dedicated to aspiring fig pigs living in coldish climates such as mine.

The mission of this book is to grow a new crop of fig pigs.

Thanks to my wife Shelley for putting up with an eccentric husband who took the basement washroom out of service every winter so he could store his fig trees there.

CONTENTS

IN THE BEGINNING: CONFESSIONS OF A FIG PIG

For me, being a fig pig started as a teenager (when my sense of right and wrong wasn't fully developed, and I pinched a cutting from a local public greenhouse). It's over 25 years later, and I still have a descendant from that tree!

In the intervening years I've interrogated fig-growing neighbours (many of whom seem to have Mediterranean roots). My casual interest in figs grew into infatuation after I worked at Read's Nursery, which has the U.K. National Collection of figs.

But after that initial teenage infatuation, my first fig tree was utterly abused for 15 years. It was rarely repotted, infrequently fed…and was a gnarled, sad-looking specimen that never fruited. But what was I to do? I was moving around with school and work and never had much more than windowsills to offer it as a home.

When Shelley and I bought our first house, I felt I could finally seriously grow figs. It was a small, 1950s bungalow. We really liked that house—and we really liked that it had a small washroom in the basement.

Any of you who have lived in an older home will know that older homes come with surprises. We didn't notice until the first fall after we moved in that the basement washroom had no heating. None. It wasn't a place you would want to hang around in the cold weather. On top of that, the window was no bigger than a toaster.

As I sat/stood in that dark, frigid little room one morning, I realized that it didn't have to be a washroom over the winter. I can still hear Shelley's surprised voice: "What on earth are you talking about?" It was clear to me that it would be a fig storage room. And so we returned to leading our lives with a single, shared washroom. (So be warned: If you become fig-obsessed and if you're in a relationship, be sure that you have an understanding partner.)

Fig Pig is a term I borrowed from the late English garden writer Christopher Lloyd, who described himself as being "passionately fond" of ripe green figs, and as an "archetypal fig pig."

Actually, I think in my case "passionately fond" is a bit of an understatement. Despite a culinarily cloistered past with Anglo-Saxon and Slavic food heritage (think Yorkshire pudding and sauerkraut), I'm crazy about fresh figs. I have a couple dozen fig trees that live with me and my family in Toronto, Canada, despite the cold, grey winters.

I thought I was an eccentric fig-growing nerd. But I realized I wasn't alone when I gave a talk about growing figs at a Seedy Saturday gardeners' event. I brought a couple of extra fig trees for the draw. The winner of one of the trees came up to me afterwards and said that she had multiple offers to buy the tree. Hmmm, I thought, maybe figs are more popular than I realized. So here it is, a book to egg on all you eager fig pigs in coldish climates.

Believe me, it's possible: In the late 1800s, fig orchards could be found on the outskirts of Paris, France—well outside boundaries dictated by the practicalities of climate. The Parisian demand for fresh figs—and the fact that ripe figs quickly spoil—made it worthwhile for farmers to go to great lengths to supply the French capital. Yes, you can supply yourself with fresh figs too!

INTRODUCING TWO OTHER FIG PIGS

In this book you'll find insights from Fig Pigs Adriano Ferreira (in North America) and Stephen Read (in the U.K.), two fig aficionados whom I greatly admire. I've invited them to share their pointers as there's more than one way to grow figs—and because these guys are a font of fig facts.

Adriano Ferreira

In Oakville, Ontario, Canada, Adriano Ferreira maintains an extensive private collection of figs, with over 200 varieties. His website states, "Our mission is to put a fig tree in every home." Whenever I visit Adriano I come home with great fig-growing ideas and a belly full of figs. I once asked how big the collection will grow, to which he replied, "I think I'm gonna stop." I'm not sure I believe him, as he has a new addition to the collection whenever I talk to him! Read the profile of Adriano at the end of this book, and visit www.adrianosfigtrees.com to learn more.

Stephen Read

In Bungay, U.K., Debbie and Stephen Read run Reads Nursery, a sixth-generation family business specializing in fruit trees—and holder of the U.K. National Collection of figs. I worked at the nursery for a delicious summer during which I grazed my way through the collection, consuming countless delicious figs. Stephen was always very understanding when I unknowingly ate figs he planned to photograph. Visit www.readsnursery.co.uk to learn more.

ABOUT FIGS

Why Grow Figs?

A fig picked too early never ripens to perfection. It's just a corky, semi-sweet thing passed off as a fig. Contrast that to a truly ripe fresh fig, which packs a succulent burst of jam-like sweetness.

That perfectly ripe fig is far too fragile to withstand long-distance shipping. And that means many people living in colder climates have never experienced one. It also means that store-bought figs passed off as fresh are rarely truly ripe.

My first exposure to figs was to the dried sort. I don't mean to disparage dried figs, but they would not be reason enough for me to go to the effort of growing figs in my garden. When people say they don't like figs, I'm immediately suspicious that they have only tried dried ones.In the garden, waiting for figs to ripen can test the patience of any hungry fig pig. But then the change happens: The neck becomes soft, the fruit droops and becomes soft to the touch, and sometimes a glistening drop of nectar escapes from the "eye" (see page 25).

Even if you have a source for ripe figs, fig trees in the garden offer something else: superb architectural qualities. No other plant combines grey, elephantine bark, exotic leaves, and the warm feel of a Mediterranean climate.

A truly ripe fresh fig packs a succulent burst of jam-like sweetness.

Someone who has tasted only dried figs might not understand the fascination with this fruit.

Growth Habit

The first thing to remember about figs is that they are very, very forgiving plants. Remember that fig cutting I told you I took as a teenager? It languished under barely survivable conditions for 15 years until I gave it a proper spot.

The next most important thing about the way figs grow is that they lose their leaves after first frost. They WANT to go dormant. That means you can keep them over the winter even if you don't have a bright, hot greenhouse. While they're dormant, they don't need light or much heat. Contrast this to lemons...

The large, lobed leaves grow on branches with grey bark. When cut, branches exude a milky white sap that can be irritating to the skin. Leaves are deciduous, which means they fall off when cold weather arrives. Unpruned, plants usually grow into a bush, but can be trained into small trees if that's what you prefer.

What we call the fig fruit is not truly a fruit in a botanical sense: It's a hollow receptacle filled with tiny flowers massed on the inside wall (a syconium if you want to be correct).

Talk of flowers brings us to pollination. In fig-growing regions, a tiny wasp, with the imposing Latin name of Blastophaga grossorum, takes care of pollination. But we cold-weather types don't worry about pollination. (And I don't care much for Latin names anyway.) To understand why, let's look at the four distinct types of figs:

1. Caprifigs: Produce small, inedible figs, but the flowers inside the fig produce pollen required for pollination and development of Smyrna and San Pedro figs (see below);

2. Smyrna figs: These produce edible fruit, and are pollinated with pollen from caprifigs by the Blastophaga wasp. When not pollinated, the figs drop prematurely. Smyrna figs are not grown by cold-climate hobbyists because we don't have the wasp to pollinate them, nor the caprifigs to provide the pollen;

3. San Pedro figs: The breba crop (see sidebar) will grow without pollination, but the main crop requires pollination with caprifig pollen. Some people will grow San Pedro figs for the breba crop alone (see Desert King, page 29); and

4. Common figs: These are the figs that cold-weather gardeners grow because they develop fruit without insect pollination.

A large, lobed fig leaf.

The middle of the stem, this year's growth, is changing from green to brown. It's the current year's growth, made in the summer, that is starting to turn woody. It's on this current year's growth that the main crop figs are borne.

A tree that recently came out of dormancy. There are already figs sprouting from the woody growth. You'll often have pea-size figs that swell and start growing out of last year's wood even before you have leaves.

Two Crops

You can look forward to two crops per year from your figs—though in some areas not all varieties ripen both crops. And some varieties, even of common figs, just aren't suited to colder areas.

The Breba crop—the first crop, the summer crop—arrives on growth from the previous season. It ripens as early as July where I live, but the timing will vary with the variety and the weather.

Some fig aficionados consider this crop to be inferior to the main crop in some varieties, and speak of it in disparaging terms. Not me: I gobble up any that grow in my garden.

Two Crops

The first crop—called the breba crop—is formed on growth from the previous season.

The second crop—the main crop—develops on new growth.

Common Fig not so Common—and is What Most Amateurs Grow

Common figs are what cold-weather gardeners grow because they develop without pollination.

Main crop figs—the second crop—form on new growth. In colder climates, this crop will likely ripen in September or October...but may not ripen at all if the weather doesn't co-operate and you have a late variety.

Before we had kids, Shelley and I always used to take a vacation in September, just around the time that the main crop figs are ready. My mom would tend my garden. I remember the very first year that I was expecting a decent harvest... and most of them arrived in September, while we were away. When we returned home, Mom said, "Your figs were sure delicious!" I was deflated.

CHOOSING A SPOT

Light

Fig trees do best with a lot of light—at least eight hours per day—for good fruit production. The good news is that they will tolerate partial shade, although that may result in less of a crop.

Temperature

When grown in the ground, fig trees will withstand temperatures as low as about -10°C (14°F). When grown in pots, where the roots don't benefit from the insulating property of the ground, they likely won't survive temperatures that cold.

See the section on overwintering for tactics on keeping fig trees warm enough during the winter months.

In North America, fig trees are hardy in USDA zones 8-11, but if you're reading this book, you're probably jealous of people in those zones.

Ground versus Containers

Fig trees grow very well in containers. The advantage of pots for backyard growers is that plants can be situated to get maximum exposure to sunlight. Use a heavy plastic pot that can be sunk into the ground, allowing roots to grow into the surrounding soil and bring home water and nutrients.

Another advantage of containers is that they restrict the tree's roots, which helps limit the fruiting and vegetative growth.

Good tree form means side branches at wide angles, as close to horizontal as possible. These branches are tied down to achieve a wider angle. After growing in this position for a few months the twine can be removed and the branches will stay in place. (If you guessed that this is not a fig tree, you're right...but this technique works for figs.)

A tree form. I grow my figs as trees, which I find more practical for overwintering.

TRAINING & PRUNING

Unpruned, fig plants become bushy, but with pruning can be formed into a small tree, fan, or cordon (see below).

Your job (if you're a perfectionist) is to build a framework for the form you want.

Pruning Tips

Wherever possible, prune to an outward facing bud, which encourages new branches to grow outwards, giving the plant an open structure. Remove crossing branches.

Make cuts back to a bud or branch so as not to leave a dead nub, which is an entry point for decay.

Tree Form

If you want a tree, prune to encourage one main stem and train it upwards. Stake plants for the first couple of years, pinching off side shoots until the main stem reaches your desired height. Then, pinch out the tip and allow four to six side branches to grow.

Once the tree form has been established, prune out crowded shoots. Occasionally prune back older shoots to a bud to promote the growth of new shoots.

As with many other fruit trees, good form means side branches at wide angles, as close to horizontal as possible. It's unlikely your side branches will grow this way naturally, so you may have to weight them down (see photo). Alternately, insert a "spacer," (e.g. a Popsicle stick) between the main stem and the branch in question to hold down the branch at a wider angle.

Fig Pigs on…Form

Stephen R. prefers to form his fig plants as bushes, saying, "Tall trees blow over!"

Adriano likes trees for pots, but in the ground, where he bends over the plant for the A-Frame technique of overwintering (see page 34), he prefers multi-stemmed bushes. Once a stem becomes too thick to easily bend over for the winter, "Just cut it off," he says.

Once the tree form has been established, prune to thin out crowded shoots, occasionally pruning back older shoots to a bud to promote the growth of new shoots.

Bush Form

There are a couple of ways to grow figs as bushes. If you're an orderly sort of person, you might want to exercise some control. If you're not, you might just let the bush grow as it pleases.

For control freaks, your goal is three to eight widely spaced shoots from the base of the plant. And you don't want just any shoots—you want the vigorous ones that exhibit strong growth. (Just like people, in the world of shoots there are slackers too.)

The multiple stems might all come from below ground, or they may come from above ground. Either way, keep the best, discard the rest. Cutting out the weaker branches every year in early spring does something other than leave you with the best: It also encourages new shoots from the base.

Remember

If you have a tendency to get carried away while pruning, remember that what you prune off in the fall will limit the breba crop of the following season because you've pruned away the old wood that bears that crop.

A fig plant grown in a bush form. With four well-spaced main branches, this bush is not too cluttered.

Below: Forming a fig plant into a bush from a young cutting. I like the look of three to eight widely spaced shoots.

My Favourite Form

I grow my figs as trees or bushes—no fans or cordons. It's not that I dislike these latter forms (in fact I love their architectural properties), but they're a bit less practical for overwintering.

Fan Form

If you're a form fanatic, then consider a fan. A fan is a plant that has one stem coming out of the ground, with a number of shoots coming off that stem, so it resembles a fan.

Fans are more common when figs are grown permanently in one location. It's difficult to bury a bulky fan-shaped plant, or to find indoor storage for it.

I recently saw a fan fig tree for sale here in Toronto for hundreds of dollars. It had a thick main trunk, nicely angled shoots, a sturdy trellis, and was in a pot nearly the size of a half barrel.

If, like me, you're not ready to fork out hundreds of dollars for such a specimen, but you want a fan, don't despair. You can do it yourself. Here's all you need to do:

1. Cut back the fan-to-be plant to a single stem, about 30 cm (12 inches) high, which will encourage the stem to send out lots of new shoots;
2. Train these new shoots to the left and right; and
3. Build a trellis onto which you can tie the shoots to hold them in place.

The following spring, cut each of these new shoots to about 30 cm (12 inches) long, to encourage new growth.

A fan is a plant with one stem coming out of the ground, from which a number of shoots veer off at angles to resemble... you guessed it... a fan.

Cordon

If you're thinking food right now, a fig cordon has nothing to do with chicken cordon blue! Think cordon as in a police cordon—a line of officers. In the case of a fig grown as a cordon, you're training it as two horizontal shoots (one to the left of the main trunk, one to the right) not far above the ground, with many vertical branches coming up from these.

I have never seen a cordon for sale—and I'm not surprised, because it would be very impractical. But you can make your own:

- Cut the main stem back to approximately 30 cm (12 inches);
- Make the cut just above a leaf axil, so that you're likely to get new growth on either side;
- Allow one shoot to grow on either side. For the first season, don't force them to grow horizontally, but let them grow at about 45 degrees. The shoots should grow more quickly at that angle than if you bend them horizontally;
- In the fall, pull down the two shoots so that they are horizontal
- The following spring, cut back each shoot to three nodes (look for leaf scars);
- Each of the nodes on the horizontal shoots should now send out a new branch upwards for you. Use stakes or wire for support.

Timing

Pruning for shape is best done in the spring, when plants are eager to make new sprouts. Some sources recommend spring pruning for established plants because this lets you remove any dieback that has occurred over the winter.

I'll be the contrarian: I prune my trees in the fall, once the trees are dormant. This allows me to fit more plants into my small storage area.

WATERING

Many people think of fig trees as drought-loving plants. It's not true. Leaves turn yellow and drop if there is insufficient water—and, worse still, developing fruit will quickly abort.

Not only do fig trees need water—they need regular watering because sporadic watering can cause fruit to split.

Whether your in-ground fig trees need water will depend on your local conditions. But if you grow figs in pots, you must plan to water them regularly, regardless of your local conditions.

When growing fig trees in pots, the easiest way to know that you've watered enough is that some water comes out of the bottom of the pot. Of course, this visual clue is no longer available once you sink your pots in the soil a bit…but in that case the roots will creep out into the ground and be able to scavenge some moisture.

Still not sure if you need to water your potted fig? Poke your finger into the pot to find out if the soil below is still moist. If the roots in the pot are too dense to allow you to poke in your finger, use a stick or something pointy to make a hole into which you can stick your finger. In high summer, in my climate, I find that actively growing figs in pots require daily watering.

Cut back your watering in the fall as the fig plants become dormant.

Sink Those Pots into the Ground

I like to sink my potted figs into the ground a couple of inches when I put them outside in the spring. That's because they'll quickly send roots out into the surrounding soil, and scavenge water and nutrients. It doesn't mean I don't have to water them, but it serves as a bit of a buffer if I'm slow.

If you do this, be prepared to chop free the pots with a spade in the fall, severing all the roots that have slithered out of the pot and into the surrounding soil.

FEEDING

When I was a newspaper delivery boy as a kid, I delivered to one house where there was a ginger cat on the front porch. This wasn't any old cat: it was the biggest, fattest cat I have ever seen. And because it was so big, it was extraordinarily slow (for a cat). That cat was not doing what I, as a gardener, would want a cat to do, which is get rid of rodents.

In the case of pets, overfeeding can undermine usefulness. It's the same with figs (although I'm not saying that fig trees are necessarily pets). If you overfeed your fig tree, it will get big...like that cat. You'll have lots of lush green leaves. And you'll have fewer (or no) fruit. So when it comes time to feed your fig trees, think of that big fat cat and make sure you're not overdoing it.

Whether to feed or not depends on whether the plant is growing in the ground or in a pot. If it's grown in a pot, supplemental nutrition is always necessary. When trees are grown in the ground, fertilizing may not be necessary if the soil is rich. I tend to err on the side of cheapness.

Not Too Late!

Skip the late meal! I dread being invited out for a late supper because the consequence for me is always the same: I eat too much, too late, and sleep poorly.

When it comes to feeding your figs, don't feed them too late in the season or you'll mess up their winter sleep. Food late in the season can encourage lots of lush, green growth, and such growth might not "harden off" for the winter. That means it's extra susceptible to cold damage. And even if it isn't damaged by cold, your tree might be a bit slow to become dormant.

Fertilizing Pots, 1-2-3

With pot-grown figs, I follow a simple, three-pronged approach to feeding:

- I use a potting mix containing garden soil, which not only contains some nutrients, but also has the capacity to hold nutrients;
- I feed with a liquid fertilizer somewhat regularly (I don't say regularly, because that would be fibbing); and
- I apply a low dose of slow release fertilizer every spring so that if I miss applying the liquid feed (which I invariably do), my plants don't starve.

Choosing Fertilizers

Those Numbers

Many sources recommend a fertilizer with a high level of potash (that means a high last number—e.g., 15-15-30) during the growing season. Strictly speaking, this recommendation makes sense...but wait a second!

I stick with an all-purpose fertilizer (one where the numbers are roughly equal—e.g. 1-1-1 or 20-20-20.) Here's why: As a home gardener, I don't see the point in having countless fertilizer formulations around the house, so I stick with one general-purpose product.

Other Considerations

- Some people include composted manure in their potting mix, which adds food for the fig plants;
- Sinking your potted figs into the ground allows roots to come out and scavenge nutrients from the garden soil; and
- Just remember that with figs, when it comes to food, less is usually more. Your job as the fig grower is to keep in check the leafy growth of the plant, and one of the best ways to do that is to keep it on the hungry side.

Tip

Stop feeding in late summer, because young (green) growth can be damaged by fall frosts if it has not been "hardened off."

Fig Pigs on...Feeding

Stephen R. prefers to top-dress with a slow-release fertilizer in the spring and again in late June—coupled with weekly sprays of a liquid seaweed fertilizer.

Adriano feeds his fig trees a slow-release seaweed fertilizer once per summer; compost tea every couple of weeks; and a liquid foliar fertilization monthly. Along with that, he has a source for rabbit manure, so will sometimes add a bit of that to the pots.

POTTING & REPOTTING

When developing a young plant for my collection, I move it to a bigger pot once or twice annually. The frequent repotting and added soil stimulates rapid, vigorous growth because it gives the roots space to grow. As plants grow bigger, they can also be moved to successively larger pots.

Once the plant has attained the desired size, I repot it (without changing pot size) every three to five years in the spring. For me, the best size is a tree about 1.5 metres (five feet) in height (which, after summer growth, will be taller), a size that is easy enough to get through a doorway.

Repotting will likely affect yield the year that it is performed, but it's important because it (a) maintains vigour, and (b) keeps the plant the size you want.

I label my trees with a metal tag that will withstand weathering (plastic tags may degrade, and marker on plastic tags will wear off.) I write the variety name and the year I repotted the plant.

Here are my steps to repotting:

- Lift the plant from the pot;
- Remove some of the old soil mix and roots—but no more than half the root ball;
- Prune back some branches to open up the canopy; and
- Do not move to a larger pot unless you want a larger plant!

One final note about potting up your trees: Don't make the soil level with the top of the pot. You want the top of the soil to be 5–10 cm (two to four inches) below the lip of the pot so there's room to water generously without overflowing.

A Fig Pig Says... Prune Roots

Adriano uses root pruning to maintain the vigour of his potted collection. "Take a saw and cut it—that's it," he explains, adding, "Cut one quarter of the root." He recommends root pruning at least every four to five years.

CONTAINERS

When it comes to containers, my preference is commercial nursery pots. That's because they're not only inexpensive: they're durable. I've tried many ornamental containers, but often they become brittle and crack after a winter or two.

I don't recommend clay pots, even though they look nice. First of all, they're heavy. Second, they're porous, and wick water away from your thirsty fig trees. If you do decide you must use a clay pot, line it with plastic, but poke few holes in the bottom so excess water will drain away.

When it comes to containers, my preference is black commercial nursery pots. They're inexpensive and durable.

Container Ideas

Here are some other ideas:

- Saw a plastic barrel in half and you have a hefty, long-lasting container. For a large container like this, you'll want to have a plan for moving it when winter arrives. (And if your storage area has stair access…maybe use a smaller container!);
- Use the large pails used in the food industry, making sure to drill holes at the bottom;
- Make a container out of wood. (Don't use pressure-treated wood.);
- If you opt for nursery pots, don't be shy about garbage picking. I've harvested lots of good pots that way and saved a bundle of money. Look for people who are having landscaping done and new shrubs and trees planted; and

A Fig Pig Says…Add Holes

Adriano says that the most important fig-growing tip he gives to first-time growers is to make additional holes in the bottom of the pots. "You have to be a strong guy to take it out in the fall," he says, describing the extra roots that grow out of the added holes into the ground. The holes also allow for improved drainage.

- Where I live, people have rectangular plastic bins that were formerly used to collect recyclables, measuring roughly 38 x 50 cm (15 x 20 inches). They're durable and a suitable size for a fig tree. Just drill a few extra holes in the bottom to improve the drainage.

Pot Size

There is no "right" pot size. It really depends on what fits your circumstances. My former colleague Frank talked about his father's fig tree, which was in a large, half-barrel-sized container. Frank's dad had installed wheels onto the bottom of the pot—wheels that could be locked once it was in place for the summer. In winter, he would simply wheel it into his garage, which was attached to his house and so stayed warm enough to overwinter the tree in a dormant state.)

I don't have an attached garage, so my potted trees must either be schlepped down the stairs into the basement, or through a doorway into the shed. That rules out wheeled, half-barrel pots for me. They're too heavy to lift.

My preferred pot size is 36 cm (14 inches) in diameter and about 30 cm (12 inches) deep. The exact measurements will vary by manufacturer, but I chose this size because, when it was filled with soil and a tree, it was the largest and heaviest that I wanted to be moving around. If you can visualize things better in terms of volume, that's 25 L (6.5 US gallons).

SOIL I like to think of soil mixes for my fig plants as either:

- Short-term mixes for rapid growth; or
- Long-term mixes for sustained growth.

By short-term mixes, I'm referring to soilless mixes that are light and fluffy (lots of air space allowing roots to grow easily)—and, when combined with ample food and water, encourage vigorous growth. This is the soil mix I use for growing the tree to its final size.

Once a fig plant has reached the desired size, I change gears and give it a heavier soil mix that will dry more slowly, hold more nutrients, and is better suited to long-term growth.

I liken mixing soil to cooking. Gardeners all have their own favourite soil mix recipes—and the recipe may change with the mood of the gardener. Some incorporate garden soil, some don't. Some people may use compost, while others won't. I like to add garden soil because it has water- and nutrient-holding properties, and I like peat because it makes the mix lighter. I also like to add sand for both drainage and weight (so the pot won't blow over).

Fig Pigs on...Soil

Stephen R. uses a John Innes No. 3 soil mix modified by including 20 per cent composted fine bark. This, he says, improves the drainage, and makes the mix lighter in weight and easier for roots to penetrate.

Not sure what John Innes No. 3 is? John Innes soil mixes (or "composts" as they are known in some parts of the world) are soil-based seed and potting mixes containing loam, sand, and peat, along with limestone and fertilizers. They're a classic in the U.K., where they were developed by the John Innes Horticultural Institute in the 1930s.

Adriano says that he sometimes adds fine bark too, for drainage. His soil mix starts with a commercial soilless mix (peat, perlite, and vermiculite based), to which he adds a sprinkle of bone meal, some rabbit manure—and will sometimes add garden soil (loam). If he has egg shells, he'll use those too.

Steve's Simple Long-Term Soil Mix

- 1 part garden soil
- 2 parts peat-based soilless mix
- 1 part sand (depending how much my kids let me take from their sandbox)

Any experienced gardener will look at my recipe and say, "So what sort of garden soil do you have...and what if mine is different?" That's a great question—and it comes back to what I said about mixing soil being a bit like cooking: If you cook well, you're probably always tasting to make sure you have things right. When you mix soil, you're not tasting (I hope you're not), but you are judging the feel of it and adjusting the proportion of ingredients accordingly.

I don't use the sticky yellow clay subsoil that I have around here. Whenever I remove lawn, I stack the sod upside down. As the grass dies and decomposes, it leaves a decent (for my area) soil with organic matter from the decomposed grass shoots and roots. That's the garden soil I use for my figs. (If soil nematodes are a problem in your area (see page 39), you won't want to use garden soil, unless it's been sterilized.)

HARVEST

One or Two Crops

Some varieties of figs will reliably give two crops annually in coldish climates, although some will give only one.

In some cases, the overwintering breba crop can be removed to encourage the formation of the main crop. On other varieties the breba crop is best—and the main crop will never ripen in time.

Fruit Drop

Even in ideal growing conditions, some fruit naturally drops before maturing. This is normal—and not the hallmark of any lack of fig-growing skills. But...allowing trees to become too dry between waterings will cause even more fruit to drop—ones that would otherwise have held on until harvest time. So remember, water well when actively growing to prevent your prized fruit from falling off.

A Fig Pig Says... Pinch Those Tips

Stephen R. says that the most important fig-growing tip he gives to first-time growers is: "Pinching the tips out in summer is easier than pruning."

Promoting Fruiting

During the summer, you can pinch out the shoot tip after four or five leaves have appeared to encourage main crop fruit formation.

Any small, unripe fruit remaining at the end of the season should be pinched off. They will not ripen, nor will they ripen in the coming year.

Ripening

Fruit will continue to ripen in late summer and early fall as long as the weather remains warm.

If the season is nearly done and there are good-sized fruit that aren't yet ripe, some growers place a drop of olive oil at the base of the fruit (over the eye) to speed up ripening.

Not all fruit ripen at the same time.

My daughter Emma is very fond of figs and loves to help me pick them. Drawing courtesy of Emma Biggs.

Picking

I try to pick my figs first thing in the morning, while all is quiet. There is no horticultural reason for this. It's because I don't want to have to fend off my daughter, Emma, who is also a fig pig, and who will insist that I share with her.

It is important not to harvest fruit too early because a fig picked too early won't ripen any further. You'll know that you've picked it too early if the stem bleeds a milky sap.

Telltale signs that a fig is ripe are:

- A soft neck, causing the fruit to droop;
- A drop of nectar coming out of the eye of the fruit;
- Slight tearing of the skin;
- The fig retains the impression of fingerprints when touched; and
- Ants or hornets feasting on the fruit!

In addition to colouring and tearing of the skin, figs usually droop when they are fully ripe. This fig has developed a nice colour, but is not yet drooping.

This fig is drooping nicely.

In many cases, the skin of a fully ripe fig will begin to tear.

Some varieties tear more than others. (Photo courtesy of Stephen Read, Reads Nursery.)

Some fig varieties have a prominent "eye" at the base of the fig. A glistening drop of nectar will sometimes come out of the eye of a fully ripe fig.

Enjoying...

Stephen Read

Stephen R. says that his favourite way to serve figs is in a steamed fig pudding.

Steamed Fig Pudding

1 cup of sugar

3 egg yolks, beaten (reserve egg whites)

1 cup of milk

1 teaspoon of vanilla

1 pound (or more) of figs

1 apple (peeled and cored)

2 teaspoons grated orange rind

1/2 teaspoon cinnamon

1/8 teaspoon ground cloves

1/8 teaspoon powdered ginger

1 1/2 cups dried bread crumbs

2 teaspoons baking powder

"All I do is mix the sugar eggs-cream together then add the other ingredients (except egg whites— whip these and fold in last.) Steam for 3-4 hours slooooooowly."

Adriano Ferreira

Adriano likes his fresh, saying, "Fresh! I don't like them with anything else."

Fig Pigs on... Fig Overdoses

I asked the Fig Pigs if they ever have too many figs to eat.

Stephen R. is a generous guy. He says, "No, as we allow some for the birds." That's a great problem to have! I'm less kind-hearted and never leave ANY figs for the birds. None.

Adriano told me, "The thing is...I have friends and family come, so there's never too many." I know, because I've been there and eaten figs while touring his garden!

OVERWINTERING

Let's get things in perspective before we begin looking at ways to overwinter fig trees in colder climates. Over the winter, your fig trees will need no more care than a potted houseplant—and probably less, because they go dormant.

I break down overwintering tactics into two broad groups: indoors and outdoors.

Easy Because Figs go Dormant

Fig trees are deciduous and drop their leaves after first frost. This is great news for gardeners because it means we can keep them over the winter even if we don't have a bright, hot greenhouse. While they're dormant, they don't need light or much heat. In areas where figs won't normally overwinter outdoors, this dormancy means they can be stored inside, or, in many cases, outdoors with minimal protection.

The key is to keep the plant in a dormant state as long as possible, by storing it in a cool, dark place, Active growth indoors, in low-light conditions, will give weak, lanky growth.

Getting Ready for Winter

With the arrival of fall, decrease watering potted plants to allow the soil to become fairly dry as plants drop their leaves and become dormant. The plants require less water as they go dormant—and sopping wet soil could cause root rot.

Get Over It

It will break your heart…but there will be figs that don't ripen. As you tuck in your trees for the winter, remove any remaining figs that are bigger than the size of a pea.

Planning to keep your figs growing all winter? I don't recommend it.

Think you're going to grow your fig in a window over the winter? Think again. Window light just doesn't compare to outdoor light, so you're likely to get lanky growth. Besides, with dormant plants you don't need to worry about bugs…while with actively growing plants you do.

And In the Spring...

In the spring, frost won't hurt your trees if the buds haven't broken (meaning leaves and breba figs haven't started to grow); but if they have, beware! A frost will probably damage them and may affect your crop. One year I moved my figs to a sunroom in the spring after they started to grow before I could put them outside. The problem was that I forgot to turn on the heat one night when the temperature dipped below freezing. Those tender new shoots were all frost-bitten the next morning…affecting my breba and main crops.

This might leave you wondering when to take your fig trees outside, or when to unearth or unwrap trees that are under cover. How's this for an ambiguous answer: It depends. It depends on whether you're willing to rush them under cover if there's a chance of frost. If you're not willing to schlepp around plants, hold off. But, of course, the sooner you get them growing, the sooner you'll have figs… and the more likely you are to have main crop figs ripening before winter returns.

Overwintering Indoors

We'll start with a look at what you can do indoors. Pot-grown fig trees can be stored indoors, in a dormant state. You simply allow the plant to become dormant (drop leaves) before bringing it indoors. You want somewhere cool and dark, so your plant doesn't decide to grow before you want it to.

If you have an attached garage, cold cellar, or unheated basement, these can all be good places to overwinter the plant. Ideally, the temperature in there should be between 7°C (45°F) and about -3°C (27°F).

The KEY is to keep the plant in a dormant state because growth indoors, in low-light conditions, will be weak and lanky. Cool temperature and low light levels—better yet, darkness—help to keep the trees in a state of dormancy as long as possible.

The dormant plants will still need a bit of water from time to time. Not too much, though—they can be easily killed by overwatering when they're in this state.

Dry Soil Tip

Soilless mixes containing a lot of peat contract when dry. This causes a problem when you water, because the water will not soak into the soil, but simply run down the gap around the inside of the pot. Before watering, push down the mix around the edges of the pot. You can also make a hole in the soil with a stick and fill it with sand—creating a channel into which water can penetrate.

Indoors in Greenhouse

I want to reiterate that you do not need a greenhouse to grow figs in your coldish climate.

But some people use them... I saw the hand-painted roadside stand and had to stop. Luckily, my friend Donna, who was travelling with me, is also a plant buff, so she didn't mind stopping for what would likely be a long visit. The sign said simply, "Fig Trees for Sale."

The owner came out, got us past the barking dog, and took us to a line of potted fig trees that he was selling. He had labelled the varieties with the names of the places from which he got them in his native Italy.

Me being me, of course, it wasn't enough to just see the trees for sale and buy a couple (which, of course, I did). I needed to know how he overwintered them. But the information wasn't forthcoming.

Finally, after a lot of questions and probing on my part, he gruffly motioned to follow him in a way that gave me the feeling he didn't do this for any old visitor. We were about to be shown the central hub of his fig operation.

In the greenhouses behind his house he had space to overwinter these pot-grown figs. But he also had in-ground trees—sprawling specimens with thick, elephantine bark-covered stems and a canopy that, in many places, made me feel as though I were on the forest floor.

He fed us fig after ripe fig (and cactus pear too) as he told us that the trees had been in the ground here for many years—and that if he were to do it again he might do it differently. He lamented having put them in the ground because they were so vigorous and needed so much pruning.

And that's the problem when you grow figs in the ground in a greenhouse: They become wild and unrestrained. Many texts prescribe making a pit, lining the sides with concrete blocks, and then putting crushed rubble at the bottom. This restrains the root system, and gives a tree that won't be all leaf and no fruit. It's almost like an in-ground pot! But, whew, that's a lot of work.

If You Use a Greenhouse

If you have a greenhouse and plant your tree in the ground, restrict the root run so that you don't get excessive growth at the expense of fruit. Dig a pit approximately 60 cm (two feet) deep and wide, and line it with upright patio stones. At the base of the pit, use a layer of broken stone or brick to allow drainage while restricting excessive root growth.

Indoors in a Shed

I no longer haul my fig trees into the unheated basement washroom for overwintering, I've found a better way. When cold weather arrives, I dig them and their pots out of the garden bed and move them over into a large shed. It's insulated and has a heater that keeps temperatures from falling too low.

Indoors in a Garage

If you have an attached garage, that could be a perfect spot for overwintering your fig. But if your garage is detached, the temperature isn't moderated by the house, so it might get too cold.

Here is a greenhouse that is kept around the freezing mark most of the winter, and then heated up for seedlings in late winter. This way fig plants start growing a bit earlier.

My shed has insulated walls and a heater that keeps it just above freezing. That's actually warmer than needed for figs—but my Brugmansia plants are also in there and don't take well to freezing.

Once buds "break" (once leaves open) in the spring, your tree is frost-sensitive. If your fig plants awaken early, you can use clear plastic to create a temporary spring greenhouse that you can remove once the risk of frost is gone.

Early Risers

Wake up a napping toddler too soon and you've got a cranky kid on your hands; wake up your dormant fig tree too soon and you've got a problem! With toddlers, you're worried about keeping things quiet. With fig trees, you don't want too much action—light, heat, or water—or they may snap out of dormancy and begin to grow.

Coming out of dormancy too early means trouble: Before the buds broke (the leaves opened), your tree was frost hardy; a frost—a drop below freezing—was no problem. Once a tree's out of dormancy, frost is a problem.

Remember, in-ground figs can take temperatures down to about -10°C (14°F). But once those buds break, your tree is now frost-sensitive. A frost wouldn't likely kill the whole tree, but it would kill all those tender green leaves, shoots, and wee figs tucked in with them.

So now you have a tree that's growing, but can't take outdoor conditions. And if you were storing your tree indoors as a dormant tree, you probably don't have a greenhouse or solarium, where you could put it now. And if you put this now-growing tree in an inadequately bright spot, the growth will be lanky.

Moral of the story? Try to keep your trees dormant as long as possible.

Watering Dormant Figs Indoors

Dormant figs are more likely to die from overwatering than from underwatering.

While plants are dormant, it's important to strike the right balance in watering: enough that the dormant trees don't dry out and die, yet not too much that the roots rot. I usually check them about once a month.

Indoors

You want to keep your plant dormant, which requires cool, dark conditions.

COOL + DARK = DORMANT

If you're not sure whether your plant should be watered, stick your finger in the soil and feel around, digging down a couple of inches to feel if it's moist.

Overwintering Outdoors

When we overwinter figs outdoors, we're taking advantage of the fact that the ground remains warmer than air—something we can use to keep our plants from suffering the extreme cold of the air temperatures.

In Doubt about Outdoors?

Not all of the outdoor methods will work in all cold climates. Instead of trying to pigeonhole individual techniques into specific growing zones, I've decided to let you do the ground work here. There are so many microclimates that there's no way to spell out all options for all areas. If in doubt, start by overwintering your fig plants indoors. Then, find people in your area growing figs outdoors and question them relentlessly. Trust me, such people exist, and now that you're thinking about figs, you'll start to notice them.

Because their roots are protected in the ground, fig plants that have been hardened off can survive temperatures down to about -10°C (14°F), compared with about -7°C (19°F) for pot-grown plants. (Ability to withstand cold temperatures, though, is not the same for every variety.)

The methods below work for fig trees, as well as bushes—with one exception: The A-frame can only be used for bush-form figs, because trees tend to be too rigid to withstand such extreme bending. Which methods are suitable in your neck of the woods will depend on how cold it gets where you live. For example, the Mummy Method (see below) isn't reliable where I live.

Outdoors – Buried

When it comes to overwintering figs outdoors, the method most people seem to know is burying them. Earlier, I mentioned fig orchards on the outskirts of Paris, France. This is the sort of technique that was used there. I call it the graveyard method.

I used to bury my lone fig tree, but this method is no longer practicable given the number of trees I have. But don't discount this method if you haven't a suitable spot indoors for dormant plants.

If you dig the hole deep enough, you can cover the plant with a layer of soil, although this is overkill in my area. A mulching material suffices here. If in doubt, deeper with more mulching and soil over top will be safer for your fig.

Here's what you do:

- Dig a trench on one side of your fig. It should be as long as the tree is tall. Where I live all that is needed is a trench just deep enough to bury it;
- Tie together the branches once your tree or bush is dormant (this is so you don't need to dig as wide a hole);
- About 30 cm (12 inches) from the trunk on the side opposite to your trench, chop down with a spade to sever the roots, which will make it easier to bend over the plant;
- Bend the plant so it lies in the trench, then weigh it down with something heavy or peg it into place;
- Fill the trench with a thick layer of mulch, and cover with a tarp (you can cap this with a board and soil for additional insulation if you like).

Burying an in-ground fig plant. Dig a trench on one side of your fig, and then on opposite side to your trench, chop down with a spade to sever the roots, which will make it easier to bend over the plant. Bend the plant, and then weigh it down with something heavy, and then cover it with a thick layer of mulch, and a tarp.

Outdoors – Above Ground (Frame Up)

If you're not up to digging holes every year, there is an easier way! In my home town is a fig grower whose beloved in-ground fig tree has a metal frame (made of one-inch metal tubing) around the outside of its canopy—and in the winter he slaps insulated walls and a roof onto the frame, and puts electrical heating coil around the base of the plant.

Outdoors – Above Ground (Mummy Method)

In areas slightly warmer than mine, but where figs are not fully hardy, some people wrap their fig trees in an insulating material along with a waterproof layer. Think of a fig mummy!

Here's an example of how it works:

- Tie the branches together, pulling them inwards so the plant isn't as spreading;
- On a dry day, once cold weather has arrived, wrap it;
- Use burlap, tied or pinned into place;
- You can then make additional insulation layers, using things such as brown paper and old carpet;
- Mound soil around the base of the plant; and
- Some people use tar paper as the final garment, avoiding plastic, which doesn't breathe.

I've heard of growers who wrap snow fencing about 30 cm (12 inches) out from the plant, filling the area between the fence and the tree with leaves or straw. They then cover the whole thing with a tarp.

Outdoors – Above Ground (Packing Crates for Pots)

I call this method the Adriano method, even though I'm sure other people have used it.

What's nice about this method is that you have the advantages of pot growing (restrained roots that can give higher-yielding trees), but don't have to schlepp the pots indoors or bury them.

Simply build rectangular wooden boxes—a bit like packing crates without tops or bottoms—and leave them permanently in your garden. Here's how:

- Lay 2"x12" boards (this is what Adriano uses) on edge and nail together;
- In the winter, lay the potted fig plants on their sides within the boxes. (You can pack in as many plants as will fit comfortably—see photo.);
- Put a sheet of plywood over the box as a lid, to provide strength so it doesn't collapse under the weight of snow;
- Add a layer or two of foam sheets for insulation; and
- Finish with a tarp, to keep out moisture.

Do I use this method? Not yet. My wife advised that she's not crazy about the look of wooden frames, so I first need to install some hedging! Or, I could make it more aesthetically pleasing by constructing the frames from rock, landscape timbers, or whatever looks appealing.

These trees are ready for winter. They'll be topped with Styrofoam, plywood, and a tarp.

The same trees in the spring, after they've been uncovered.

Outdoors – Above Ground (Quickie Greenhouse)

My late neighbour Santos had both a lemon and fig tree growing in the ground adjacent to his patio, over which he had an arbour.

With the arrival of winter weather, he would put clear plastic around the patio and his prized lemon and fig trees. Then he heated the space to keep it above freezing.

If you want an in-ground plant, but don't care to bury it, consider an A-frame. Remove large branches that are too thick to bend. Shown here are branches bent at the ground.

An A-frame seen from the other end, with the tips of the branches poking out.

Outdoors – Above Ground (A-Frame)

So at this point you might be saying, "I want to plant my fig in the ground—not a pot—but there's no way I'm heating it and I'm not burying it." Here's the answer: an A-frame over a bushy fig. It's easier than you think:

In the background is an in-ground fig, with lots of pots in front. They're all against the south-facing wall of a brick house.

1. Remove large branches that are too thick to bend, which will encourage the growth of young, flexible ones in the spring;
2. Bend the tree over as far as possible. The goal is to have the remaining branches as near to the ground as possible, with nothing in between;
3. Cover the plant with insulating material such as carpet, blankets, or foam; and
4. Finish by rigging a wooden A-frame over the whole thing, to protect from wind and heavy snow.

Outdoors – Above Ground (Regrow)

My neighbour Joe had a fig beside the foundation of his house. He didn't give it any special treatment. In the winter it died to the ground, but in the summer it would regrow. If you choose a suitable variety, you could use this approach and forego the breba crop, aiming only for the main crop.

I recently met a gardener who cuts his in-ground plants to about 30 cm (one foot) in height every fall, then covers them with a thick layer of mulch. He was pleased to report that in the second year he was already enjoying main-crop figs.

Hint

Don't despair: Cold-damaged fig plants often resprout from the ground. Just be patient (and don't overwater).

A Figatorium?

My long-term goal is a figatorium. The idea—and the word—came to me from Peter Vicano, an avid fig hobbyist who jokingly calls his greenhouse with a retractable roof a "figatorium." His figatorium is mainly planted with in-ground trees, but he has a smattering in pots, too. See page 51 to read more about Peter.

The idea behind it is simple: Year-long greenhouse-growing isn't practical, nor is it ideal for fig trees. But greenhouse-like conditions in early spring (to give the trees a head start without risk of frost) and in the fall (when there is rarely enough heat to ripen all the main-crop figs) would boost productivity.

Don't confuse this greenhouse with a tropical paradise: In the winter, it's cool, so that the fig trees stay dormant.

PROPAGATING

The good news is that figs are easy to propagate. I propagate figs using cuttings and suckers...but there are other ways, too, as you'll see below.

Unless you like uncertainty, the one way you won't propagate figs is by seed. That's because figs don't "come true" from seed. Just like an apple grown from seed, the offspring will be different from the parent—so we have to clone them.

Suckers

Fig trees "sucker" profusely. Suckers are shoots that the plant puts out at its base. Sometimes they sprout above ground level, sometimes below. Those that come from below the soil level will often form roots, making it possible to simply cut them off and pot them up, with roots already in place.

When suckers form above soil level, make a raised soil collar around the sucker to encourage it to grow roots. This collar can be made with a can or plastic pot (ends removed and slit it lengthwise so that it can wrap around the stem of the plant). Then fill the collar with soil. When many roots fill it, cut the sucker off the tree, remove it from the collar, and plant it.

Alternatively, cut off the sucker and treat it as a cutting (see below). Of course, if you don't want more plants, cut the suckers off fig trees (not bushes), so they don't weaken the parent tree. Maybe you have a fig pig friend who could use them!

Cuttings

This is my preferred method of fig propagation, as I can't bear to discard fig shoots when I prune my fig trees. It's also the first way that I ever propagated a fig as a teenager.

From Dormant Plants

As I write this in winter, I have a plastic bag containing fig cuttings in my beer fridge in the basement. These are shoots that I removed in late fall, when I trimmed off a few branches from my dormant fig trees to fit them into storage. I shrouded them in damp paper towels, before placing them in the plastic bag.

From Actively Growing Plants

While some growers use dormant branches collected over the winter, others use actively growing summer wood. Most sources recommend semi-woody cuttings (meaning wood that isn't very soft and green, but has hardened a bit), though I'll root green ones if I've just snipped off a green side shoot.

Place the cuttings in bright, indirect light, in a light potting mix (I use a peat-based mix) and keep well watered. Keep in mind that you've just stressed the heck out of this little piece of branch by removing it from its parent plant. To help it recover, put a clear plastic bag over it, which will keep humidity levels high around the cutting and help prevent it from drying out. I root cuttings this way under lights in the house; but have also done it outside under some bushes when temperatures are warm.

Most of the cuttings I take are summer cuttings, which I trim to about 15 cm (six inches) long, then insert into a 10 cm (four inch) pot filled with a peat-based soilless potting mix. (A shorter cutting will often suffice if that's all you have.)

It's possible to make a cutting from both the tip and lower portion of a branch. Just make sure there are enough nodes (the places where leaves sprout from the stem, and from which new roots will arise). While you can leave a leaf on the cutting, this isn't necessary, as leaves will regrow from a rooted cutting. If you've left on a leaf and it falls off, don't worry—more will grow if the cutting roots and starts to grow.

How long this takes depends on what the conditions are. If you put your cuttings in a toasty warm spot, they'll root more quickly, of course. When in doubt, pick up the pot and look at the holes in the bottom to see if there are any wee white root tips poking out.

Smoke and Mirrors at Customs

I once met a non-smoker, a keen fig collector, who explained that he always came home from foreign trips with a cigarette pack—that was full of tiny fig cuttings.

Supplies for cuttings

To root your own fig cuttings, all you need is a cutting, soilless mix, a pot, and a clear plastic bag. Maintain a high humidity level by placing the plastic bag over the cutting. I don't use rooting hormone—although I know some fig growers do. As with any cutting, gentle heat from underneath increases the chance of success If you have a heat mat (basically, a heating pad for plants), it helps to put this under the cuttings.

Fig plants are easy to propagate by cuttings. How long cuttings take to form roots depends on the conditions.

Layering

Layering is a technique where you bring a branch in contact with the soil, stimulating the branch to make roots.

Obviously, layering is easiest if you have a fig plant with a bush form. Just bend a branch as low as possible so that it contacts the ground (or the soil in another pot that you've placed adjacent to your bush). Keep the bent branch in place with a heavy object like a rock.

To increase your chances of success, make a diagonal cut part-way through the branch (not all the way through) at the point where it touches the soil. (This step isn't strictly necessary, however.)

Allow about one month for roots to start forming. Once roots have formed, you can cut the new plant away from the parent. Place it in a partially shaded spot until it has a well-developed root system.

Air Layering

In this method, you root a branch that is still attached to the plant, but well above soil level. Choose a branch that's neither too green nor too woody—about 0.5 to 1 cm (¼ to ½ inch) wide is about right. Make a cut on an angle, about 1 cm (½ inch) long, but don't cut completely through the branch. Hold it open, being careful not to snap off the branch, and insert a toothpick to keep the cut open. Wrap wet sphagnum peat moss all around the branch where it's cut, then wrap with plastic, sealing the ends with string or tape.

Layering is a technique where you bring a branch in contact with the soil, stimulating the branch to make roots.

There are purpose-made kits for air layering, consisting of a wrap-around, plastic, pot-like appendage in which soil mix can be placed. The advantage of such a gadget is that it's easier to keep the potting mix in place and sufficiently damp. If you're a do-it-yourself sort of person, then you might wish to make your own, using a can with the bottom cut out.

PESTS & PROBLEMS

Fruit-Thieving Wildlife

It's hard to blame them…but birds and squirrels like figs, too. I've heard of fig growers who highly recommend having a cat.

Hornets will quickly move in and contentedly feed once they detect ripe figs. I simply shoo them away and pick the fig for myself.

Bark-Gnawing Wildlife

Whether you overwinter your plants indoors, or outdoors, covered, remember that to a mouse, the bark of a dormant fig tree is a sweet, sweet treat. To make matters even better for the mouse, a fig tree overwintered outdoors in a cold climate usually has straw or other good nesting material insulating it.

Ensure that you have good rodent control.

Root-Ravaging Wildlife

We're talking about nematodes: tiny, microscopic, worm-like critters that cause root swelling and tiny knots on roots. The reason that the root damage is unwelcome is that it retards the growth of your fig trees.

Nematodes especially like sandy soil, so that's where you're most likely to get them, whether on figs or other crops. (I live in an area with clay-based soil and have never had a problem with nematodes.)

Commercial growers might treat their soil with a nematicide—something to kill nematodes. As a homeowner, however, you probably can't buy such a product.

So if you do live in an area with nematodes, what's the solution? Simple: grow your figs in containers.

Good & Bad Nematodes

Not all nematodes are bad. In the garden, there are nematodes we like, too. For example, there are nematodes you can use to control fungus gnats on potted plants. The nematodes kill the larvae (immature, soil-dwelling stage) of the fungus gnats. Beneficial nematodes also kill grubs.

Fig Rust

When we get a year when the weather is conducive to fig rust, my trees can look terrible! But it's a fact of life—just like a bad hair day for people.

Fig rust is a fungal disease that attacks the leaves, appearing as small, yellow-green spots that develop into larger yellow-brown patches. These can become bigger and more numerous as the season progresses, making the plant look ragged.

In severe cases rust can cause complete defoliation, though more likely it will simply cause some leaves to drop prematurely.

While commercial growers control rust with copper sprays (e.g., Bordeaux mixture), in the home garden I'd suggest that it's not worth the bother.

Fig rust is a fungal disease that attacks the leaves, appearing as small, yellow-green spots developing into larger yellow-brown patches on the leaves.

Fig Mosaic Virus

Fig mosaic is caused by a virus that affects leaf pigmentation, causing mottled, mosaic-like patterns on the leaf. The disease is spread mechanically, meaning carried on stuff like tools or insects that have come into contact with infected trees. There is no cure for fig mosaic virus.

Don't be a fig-mosaic hypochondriac! While some commercial production guides prescribe destroying affected trees, remember that for you, as a hobbyist, the disease is of little consequence. (Even if it were, you probably don't have a lab to screen for it!)

FIG VARIETIES

There are hundreds of varieties of figs—and in some cases, more than one name for a given variety. What's a good variety in one place may not be a good variety in another—so instead of trying to cover each and every variety, let's look at some factors to consider when choosing varieties.

Fruit Colour

The coloration of fig fruit varies greatly, anywhere from light green through to deep purple, with shades of red and pink in between. The colour of both the skin and the pulp inside can vary. You can broadly think of fruit colour by breaking it into four groups.

- Light-skinned with light pulp;
- Light-skinned with dark pulp;
- Dark-skinned with light pulp; and
- Dark skinned with dark pulp.

Figs come with light pulp, dark pulp, light skins, and dark skins. (Photo courtesy of Stephen Read, Reads Nursery.)

Along with colour, the form of the fruit is quite varied too. Pictured here is a young breba fruit of a variety called Melanzana. It grows to be very long and becomes dark in colour.

Fruit Form

Along with colour, the form of the fruit is quite varied, too. Some varieties have a neck, while others have no neck. Some are pear shaped, some look like inverted cones, and some are egg shaped.

Leaves

Leaf form can also vary greatly. It is easiest to think of the different shapes by comparing them to other plants:

- Oak leaf;
- Okra leaf;
- Grape leaf (the kind you see strategically placed on statues; and
- Maple leaf.

Variety Names

Some varieties have more than one name. I grow Verte, which is also sold under the name Green Ischia.

In addition to recognized varieties, there are lots of plants you'll find without a variety name or for which someone just made up a name. A while ago I found someone selling a dark fig he brought from Calabria in Italy, which he simply called Black Calabria.

When you're choosing varieties, remember to seek out local fig fetishists and ask what does well in your area. I love the leaf shape of a variety called Brunswick, but a couple of people have told me that it's not a terribly fruitful choice here in the Toronto area.

Fruit Timing

I have one variety that is consistently far more prolific when it comes to the breba crop. Even though the main crop figs of this variety are just so-so, I love having this variety for the breba figs. If you have space for more than one tree, selecting one variety for a breba crop and another for the main crop is a good idea.

Not Just Fruit: Growth Habit Differs Too

It's not just fruit that differs between varieties: growth habit varies, too. I grow one variety that's really vigorous and puts on more vegetative growth than I'd like in a season. But it has great fruit.

Winter hardiness varies too. That means, when you're shopping around for a fig tree, track down fellow fig enthusiasts nearby, not in a far-off place.

Don't Fret About Named Varieties

I teased my friend Bob about being an orchid snob. He loves growing orchids, so I offered him one that I was given as a gift. "What's the variety name?" he enquired. When I said that it was unnamed, he was aghast and would have nothing to do with it.

Don't be a fig snob. I stupidly bought a fig from a nursery, and saw that the tag said "Higo," which I assured myself was the variety name. I speak French, not Spanish, so of course I had no idea that "higo" simply means fig in Spanish. All I was buying was a fig with a label that said fig in Spanish. Oops! I was mortified to have this unnamed fig in my midst…yet over time it's proven to be one of my most fruitful trees. So I've decided not to be a fig snob.

If you're keener than I am, you can review published scientific literature describing fruit and leaves and growth habit so you can bone up on variety traits. Me, I just want to eat figs!

Ease of Growing in Pots

If you're thinking of growing figs in pots, here is what our Fig Pigs consider the top varieties for growing in pots.

Stephen R.

Violette Sollies. A dark blue-purple fruit with a grey tint on the skin.

Dalmatie. A large green-skinned fig. The leaves are deeply "cut" and highly ornamental.

Rouge de Bordeau. A medium sized, red-flesh fruit with purple skin.

Adriano

Lattarula. A medium, light-green fig with a good breba crop. "It's a super fig!"

Celeste. Small fig with light brown skin, turning purple when ripe, and strawberry coloured pulp.

Desert King (a.k.a. King): "The best to grow in a northern climate," says Adriano of this fig. Medium to large green-yellow fig with strawberry pulp and a large breba crop. Does not set main crop in areas with no pollinators.

Hardy Chicago. Small to medium red-brown fig with strawberry coloured pulp. Known for its winter hardiness.

Violette Sollies, one of Stephen Read's favourite varieties of figs for growing in pots. It has a dark blue-purple fruit with a grey tinge on the skin.

Rouge de Bordeau, one of Stephen Read's favourite varieties of figs for growing in pots. A medium sized, red-flesh fruit with purple skin.

* Photos pages 44-45 Stephen Read, Reads Nursery.

Unusual Fruit

Looking for unusual figs? Here are the ones our Fig Pigs recommend.

Stephen R.

Orphan. "A great golden-honey coloured flesh, skin is green to very slight purple flush when grown outside here."

LSU Purple. A purple fig developed by Louisiana State University (hence the LSU in the name.)

St. Johns. A large pear-shaped fruit with pale green, juicy flesh.

Adriano

Panache. A small-medium fig with yellow stripes on the green skin.

Bianchetta. A small white prolific fig on a compact plant.

Rancona. A white fig with redish-bluish stripes.

Ronde de Bordeau, Stephen Read's favourite work-horse fig variety.

Orphan, one of Stephen Read's favourite varieties of figs for unusual fruit.

Panache, one of Adriano's favourite varieties of fig for unusual fruit, having yellow stripes on green skin.

Work Horse

If you're looking for one work-horse fig plant, here are a couple of ideas.

Stephen R.

Ronde de Bordeau. Shiny black fruit with a dark sweet flesh.

Adriano

Celeste. See above.

Overwintering Outdoors

If you plan to overwinter your fig plants outdoors, instead of growing in pots, here are varieties with which our Fig Pigs have had good success.

Stephen R.

Brown Turkey. Brown fruit with purple tinges.

Castle Kennedy. Very large fruit with greenish-yellow skin and light flesh with some red.

Excel: A light fig with an excellent flavour.

Adriano

Dessert King. "You can't beat the King in Canada!" See above.

Capelas. Not the best in pots, but one of the best in soil when bent to the ground to overwinter. Breba is huge.

Lattarula. See above.

Castle Kennedy, one of Stephen Read's favourite varieties of figs for overwintering outdoors. (Photo courtesy of Stephen Read, Reads Nursery.)

SHOPPING FOR FIGS

If you're at a nursery, you will likely be buying potted plants. But if you're ordering by mail or online, you might get either potted plants or bare-root plants. Potted plants are often shipped from spring to fall, but bare root plants are usually shipped before plants leaf out in the spring.

If the bare-root plant has a substantial root system, don't bother trimming the branches. But if it seems to have far more branches than roots, trim back the branches (and save what you trim to root as cuttings).

Q+A

"My fig froze to the ground during a snap freeze. What should I do?"

There's a good chance it will regrow from the roots. Leave it and see what happens.

"I read somewhere that caprification has to take place for figs to grow. Is that true?"

The sex life of figs really is fascinating, and caprification is part of it. But where we grow figs, we don't need to worry about caprification (which has to do with wasps and pollination). That's because we grow common figs, and they form fruit without pollination.

"Can I overwinter my fig at room temperature?"

I don't recommend doing that because it won't go dormant. Dormancy encourages good growth and fruit set. On top of that, when plants are dormant you don't need to worry about pests!

"The figs are dropping from the tree before they're ripe. What's happening?"

It's not uncommon for many fruit trees to shed some immature fruit...a natural thinning process. What you don't want is to cause even more figs to drop because you let your potted fig tree get too dry. So water regularly!

FUN WITH FIGS

Do the Fig

Sonia Day's book *The Untamed Garden* has chapters that explore our love affair with plants. They follow the path of many relationships: innocence, flirtation, romance, anticipation, deception, seduction, desire, lust, denial, passion, rapture, and devotion. There's sex too. Day advises, "The truth is, the plant world is drenched in sex. Passionate, urgent, unabashed sex."

I was happy to find the fig as the lead character in the chapter about desire, with a full-frontal picture of the fleshy folds inside the fig. Day talks about the biblical associations of the fig, but did you know that in some Mediterranean countries—those with hot-blooded males, says Day—there is a rude hand gesture referred to as "do the fig."

CONCLUSION

You can succeed growing figs: they're vigorous and forgiving plants.

More than that, figs do something that's important for growers in coldish climates: they go dormant. That means we can put them in cold, dark basements, stow them in sheds and garages, or even do them the indignity of burying them.

Remember: Pick them alone and you don't have to share them.

Biggs on Figs

APPENDIX: PROFILES OF FIG PIGS

If you've made it this far through the book, I expect you're a fig pig or an aspiring fig pig. Enjoy the profiles of Adriano Ferreira (one of the Fig Pigs whose growing tips I share in this book) and Peter Vicano, two fig lovers whom I respect greatly, and whose stories I hope will inspire you when it comes to growing figs. I originally wrote these profiles for the food magazine Edible Toronto, hoping to inspire a new crop of enthusiasts.

Adriano Ferreira Shares Fig Secrets

WHEN I'M ABLE TO BLURT OUT only a hesitant good afternoon, Adriano Ferreira pats me on the shoulder and graciously explains that the correct phrase is bom dia, which means good morning. With a mere halfdozen Portuguese words in my vocabulary, I'm far from fluent, but have eagerly scanned my phrasebook before this visit because Ferreira had told me he will only share his most important secrets if I speak to him in Portuguese.

Although we're in a Toronto suburb, you'd hardly know it in this manicured backyard filled with fig trees, plum trees, hardy kiwis, and grape vines. Ferreira points out a recent addition—a peach tree to which he has grafted a variety that bears doughnut-shaped fruit. Then, pointing to a fig tree in the ground with a trunk the diameter of a coffee mug, he says, "Anybody who has a house can have a big plant like that." And the secrets to doing so are exactly what I'm here to learn. It's figs that occupy most of Ferreira's yard—and his heart. "Once you grow them, they're the only fruit you'll want to grow," he exclaims. While he jokes about guarding his figgrowing secrets, he is very generous with his time, taking a couple of hours to coach me through growing, overwintering, pruning, propagating, and ripening figs. It's clearly a passion, and his enthusiasm shines through when he points out the techniques he's developed to overwinter and coax fruit from his trees here in Ontario, Canada.

Now retired, Ferreira started growing figs about thirty years ago while working in real estate. Reading a gardening magazine, he stumbled across the name of a fig-tree nursery in Massachusetts and noted that it was close to where his brother lived. So on his next visit out there, he contacted the nursery owner who, it turns out, grew up near Ferreira in a neighbouring town in the Azores. They struck up a friendship that helped kickstart Ferreira's fig tree collection.

Today, Ferreira's collection takes up a good part of his yard where, along with well-known varieties from figgrowing regions in California, Portugal, Spain, France and Italy, he has varieties from places such as the Canary Islands, Majorca, Austria, the former Yugoslavia, Iraq and Iran. He has two hundred varieties, which he claims is one of the largest private collections in the world—and the largest in Canada. A recent addition is a tree that came up from seed in his garden around the time his granddaughter was born; he named it after her, calling it Angelika. Ferreira fields enquiries from fig enthusiasts worldwide, both private collectors and researchers. A collector from Thailand recently paid him a visit. On another occasion, a foreign collector called just to make sure he really existed, surprised that such an avid fig collector—and fig trees—existed in Canada. But Ferreira is for real; he knows this fruit and his collection intimately, pointing to a particular tree and then describing its growth habit and fruiting characteristics.

Why figs?

Someone who has tasted only dried figs or storebought imported "fresh" figs might not understand the fascination with this fruit. Store-bought figs are never truly ripe because a fig picked too early never ripens to perfection, remaining corky in comparison to its treeripened cousin. When properly ripened, the neck of the fruit becomes limp and the body droops. Then a glistening drop of nectar escapes from the eye. This is the only time to pick it. By now, the fruit is dead ripe and far too fragile to withstand shipping.

There are light- and dark-fleshed varieties, and light- and darkskinned ones, too. Ferreira's website offers vivid descriptions such as lemonyellow skin, melting in your mouth, strawberry flavour, and honey surrounding the pulp. When I wonder about the taste difference between varieties, he likens it to wine. An uninitiated palate might not perceive subtle taste nuances, but a connoisseur will.

"That's a hard one," he says, pausing when I ask about his favourite. He especially likes a variety called Desert King because its first crop ripens earlier than most varieties. But there are others he favours because of their flavour.

Adriano Ferreira said he would only share his most important secrets if I spoke to him in Portuguese!

Growing figs in Ontario, Canada

In warmer climates, figs grow to be large, branching trees. Small, specialized pollinators—fig wasps—look after pollination in fig-growing regions. But there are some varieties that are selfpollinating (requiring no wasps) that produce fruit here.

There are two crops per season, one formed on the previous year's growth, ripening in mid-summer, and one formed on the current season's growth, ripening in September and October. The key to getting both crops is extending our relatively short season at both ends, and protecting trees from frost. "I'm always prepared," says Ferreira as he points to the plastic he keeps ready to cover plants if there's a late or early frost.

Ferreira explains the good news: you don't need a greenhouse to grow figs here. Many people with family ties to figgrowing regions keep potted trees that they bury outdoors or overwinter indoors. Instead of burying them, Ferreira bends over his inground trees and then covers them with an insulated A-frame, which offers sufficient protection from the cold. But most of his plants are in pots. These he lays on their sides within a wooden box that he covers with insulation, moderating the temperature sufficiently that they survive. On my visit in early May, they were beginning to leaf out.

Trees defoliate and go dormant with the first frost, he explains. In that dormant state, they can also be stored in a cool, dark basement. The key, he says, is to provide that dormant period. "They have to sleep," he cautions; otherwise, they might not fruit.

Ferreira's secrets

The real secret to coaxing the fruit to ripen in our climate is to gain a few days of ripening time. Ferreira shows me a couple of trees over which he's draped clear plastic bags. This creates a warm microclimate around the tree, helping it to come out of dormancy more quickly. Once the current year's growth is underway and figs are forming, another trick is to break off the tip of the branch, leaving four leaves on the current year's growth.

What's Ferreira's big secret? Extra virgin olive oil. In the first week of September, he looks for figs that don't seem as if they will ripen before winter, and puts a drop of extra virgin olive oil on the eye. After six or seven days, he repeats the step. While this doesn't work on all of the fruit, he says, it helps some to ripen.

When I ask how big the collection will grow, Ferreira says, "I think I'm gonna stop," saying he can only fit so many plants into a suburban yard. But then we visit his potting shed where he shows me a new method he's developed to root cuttings from other collectors, and he shows me how he grafts hard-to-root varieties. I'm not certain I believe he will ever stop.

Before I leave, Ferreira introduces me to his neighbour Domenic, who has an immaculate vegetable garden and fig trees in his greenhouse. Grinning, he motions to Ferreira to look at his welldeveloped fig trees. The secret, Domenic says, is that he sings to his figs in Italian. Ferreira laughs and responds that he sings to his figs in Portuguese.

Peter Vicano Builds a Figatorium

I've come to Brantford, Ontario to see Peter Vicano's fig collection, housed in the "figatorium" he built at his house. Meeting him at the offices of his construction company, we start by touring the small greenhouse he also built there, so he can have figs at work too. "I have a fetish for figs," he laughs.

You haven't lived until you've eaten a fig off a tree," he says. For Vicano, no other fruit compares to a fig. "It's the most exotic fruit, it's the only word I can use to describe a fig, exotic," he says. While Vicano is passionate about figs, he's fussy too: "You can't buy a really good fig," he says, explaining that a ripe fig should be eaten within three hours of picking.

For Vicano, besides being exotic, figs play an important role in family tradition. He tells me how it is customary to place fig leaves on the bottom of a basket filled with figs, which, when given as a gift, wishes wellbeing. Vicano says he has been playing with figs for the past 25 years, since he got his first tree from his dad, who brought a fig from Italy when he immigrated in 1956.

Figs in the Family

He describes his late Uncle Rocco as a top-notch fig grower, with whom he had a friendly rivalry. When his uncle kept sneaking into Vicano's greenhouse to inspect the crop (and enjoy the occasional fig), Vicano playfully put up a sign in Italian, entrata proibita Uncle Rocco. It's fitting that Vicano is now custodian of one of his late uncle's prized fig trees.

Vicano's aunt still grows a potted fig tree, which he describes as "amazing." He hasn't been able to replicate the large crop she gets—and she won't tell him her secret. At the age of 80, Vicano's mom, too, still tends a couple fig trees. And his four children all love figs, he says.

"A fig tree will grow anywhere, you stick it in the ground, and it will root," explains Vicano. Vicano currently has 94 fig trees from places including Northern Italy, Lake Cuomo, Capri, Sicily, and Sorrento. Despite his zeal for figs, he plays down his knowledge, saying, "I'm far from the expert." Talking about relatives who get consistently good crops, Vicano says, "Certain people have knacks." Then he suggests talking to a collector by the name of Adriano Ferreira.

Peter Vicano's figatorium, which is a greenhouse-like structure with permanent walls and a removable roof.

A Figatorium

In the past Vicano overwintered trees by burying them. But he always wondered if there was a better way. "I lay awake at night thinking about this," he says, as we talk about what he jokingly refers to as his figatorium. Vicano says many people lose the first crop of figs when they uncover their trees too early.

He describes the figatorium, in its third year, as being like a stadium with a retractable roof. The purpose is to get two pickings—and to exclude squirrels and chipmunks. Heated by a water boiler, the structure is a greenhouse with permanent walls and a removable plastic roof that comes off in the summer. "My wife really didn't want me to do it," chuckles Vicano, as we drive up and see the 84' long building in front of his house.

Figs won't taste right if grown indoors, says Vicano, which is why he has the removable plastic top, allowing them exposure to the elements. With the figatorium, he says, he eats figs from the beginning of June until November. While the picking window is only about two weeks per plant, he gets a fairly continuous supply thanks to the variety of plants. It's the second crop—arriving in September and October—which he says has a more intense, sweet flavour.

While the squirrels and chipmunks get fewer figs, there is still the occasional hitch. He explains how the heating system failed recently—and how sometimes, it gets too hot so that the figs don't get enough dormancy to fruit properly.

"I really enjoy it," says Vicano, talking about all the work that goes into raising figs. He often sends guests home with plants—and I'm no exception, returning home with two new fig trees for my garden.

INDEX

OTHER NO GUFF PRESS TITLES

No Guff Vegetable Gardening

A fun, slightly irreverent, graphic-novel-meets-gardening approach to vegetable gardening that banishes the idea that growing vegetables is exotic or difficult.

"*Your book changed my life,*" declared a reader who was empowered by the down-to-earth, practical approaches to gardening in the book.

160 pages. ISBN: 978-0-9868144-0-2.

To sneak a peek at this Canadian Bestseller, visit www.GardenCoachesChat.com.

Visit us at www.NoGuffPress.com

NO GUFF PRESS™

Book Design and layout crafted by:

FEEDLOT STUDIOS

www.feedlotstudios.ca

Feedlot Studios is all about promoting and sustaining the arts—both visual and graphic. Their focus is on educating the community by providing workshops and classes. Activities range from ceramic classes for children to web, print and logo design.

Ceramics + Web and Print Design

FEED LOT studios

ABOUT THE AUTHOR

When Steven Biggs and his wife Shelley bought their first house, he was excited to have a garden in which he could finally grow figs. The house was a small, 1950s bungalow—and, with a small second washroom in the basement, it was a big step up in the world for a couple that had, until then, shared a single washroom.

Anyone who has lived in an older home knows they come with surprises. The surprise for Biggs and his wife came the first autumn, when they realized the cold basement washroom had no heating. Sitting/standing in that dark, frigid little room one morning, Biggs realized that it didn't have to be a washroom over the winter. "What on earth are you talking about?" groaned Shelley, as he explained that the second washroom would be a storage area for his fig trees over the winter. So they went back to life with a single, shared washroom.

Biggs is an award-winning journalist and author specializing in gardening, farming, and food production. As a life-long gardener, he has gardened wherever he's lived—creating allotment gardens, container gardens, indoor gardens, and gardens in the overgrown backyards of rental houses.

Along with studying horticultural science at the University of Guelph (Ontario, Canada) he has worked in greenhouse and nursery production, plant propagation, biological controls, horticultural supplies, and farm marketing. In the garden, he favours a practical, no-nonsense—and fun—approach to things. He runs the how-to website The-Locavores-Garden.com. His book *No Guff Vegetable Gardening*, co-authored with Donna Balzer, is a Canadian bestseller.

Biggs lives with his family and a couple dozen fig trees in Toronto, Canada.